Computational Fluid Dynamics (CFD) Simulation of a Gas-Solid Fluidized Bed. Residence Time Validation Study

Baru Debtera

Bibliographic information published by the German National Library:

The German National Library lists this publication in the National Bibliography; detailed bibliographic data are available on the Internet at http://dnb.dnb.de.

ISBN: 9783346547736
This book is also available as an ebook.

Print and binding: Books on Demand GmbH, Norderstedt, Germany
Printed on acid-free paper from responsible sources.

GRIN web shop: https://www.grin.com/document/1151429

Computational Fluid Dynamics (CFD) simulation of a Gas-Solid Fluidized Bed: Residence Time validation study

BARU DEBTERA

Addis Ababa Science and Technology University
College of Biological and Chemical Engineering,
Chemical Engineering Department

Abstract

In this study, numerical simulations of a gas-solid fluidized bed reactor involving a two-fluid Eulerian multiphase model and incorporating the Kinetic Theory of Granular Flow (KTGF) for the solids phase have been performed using a commercial Computational Fluid Dynamics (CFD) software. The fluidized bed setup consists of 1.5 m height and 0.2 m diameter in which a series of experiments were performed using Helium tracer to determine the Residence Time Distribution (RTD) at various normalized velocities i.e., with different degrees of gas-solids mixing. Both 2D and 3D simulations of the fluidized bed reactor are performed. The main purpose of this study is to understand the hydrodynamic behavior of a gas-solid fluidized bed reactor through a framework of Eulerian multiphase model and to analyze hydrodynamic behavior of the gas-solids mixing. As a first approach, the CFD model is validated using the experimental results of the residence time study. The numerical results of RTD corresponded well with the experimental findings indicating that the CFD model can be used to predict the performance of the fluidized bed reactor.

Keywords: Computational Fluid Dynamics, fluidized bed, residence time distribution, gas-solids mixing, turbulence

1 INTRODUCTION

Fluidization is the operation by which solid particles change into a fluid-like state through suspension in a gas or liquid. The process of fluidization can be described basically as supplying a flow of gas through a bed of granular material at an adequate velocity such that the granular bed acts as a fluid (Gidaspow, 1994; Richardson et al., 2008). Fluidized beds are common in industry and are used for catalytic reactions, granulation, particle coating processes, heating/cooling, drying, mixing, (Kunii and Levenspiel, 1991). Gas–solid fluidized beds are advantageous for many processes involving heat and/or mass transfer between phases. They provide efficient mixing, which results in excellent gas–solid contacting and relatively uniform temperature/concentration profiles within the bed (Cui and Grace 2007; Gidaspow 1994). Indeed, the need for cleaner and sustainable energy source has led to the development of biomass gasifiers which employs fluidized bed technology, as a promising approach, given its rapid biomass heating, effective heat and mass transfer and uniform reaction temperature(Salaices et al., 2010). Understandably, the physics behind fluidization indicates the importance of variables such as pressure drop, minimum fluidization velocity, solid volume fraction profile, particle velocity profile (Benzarti et al., 2012; Taghipour et al., 2005). Usually, fluidized bed reactors are chaotic in nature (Li et al., 2009). This is indicated by a turbulent fluidized bed. In addition, a fluidized bed is a medium to carry out a chemical reaction involving gas and solid. The main reason for choosing the fluidized bed for synthesis of solid catalyzed gas phase reactions is the demand for better temperature control of the reaction zone, and the conditions in fluidized bed reactors being near isothermal.

Nowadays, Computational Fluid Dynamics (CFD) has become a powerful tool for understanding the complex phenomena between gas and solid particle phases in the fluidized bed (Hartge et al., 2009; Wang et al., 2010; Xue et al., 2011). CFD is based on solving Navier-Stokes equations and has become a powerful tool for research and development in multiphase flow systems. Computational resources have increased substantially at sharply decreasing costs, allowing the acquisition of detailed computational information about reactions and flows at a fraction of the cost of the corresponding experiments (Dutta et al., 2010). It is possible to define a computational domain in which the geometry of the fluidized bed reactor can be incorporated and CFD approach can be used to solve the governing Navier-Stokes equations, using appropriate initial and boundary conditions. The results obtained from the numerical model can then be compared to experimental data for a necessary validation. Since computational assets have expanded significantly at forcefully diminishing costs, specified computational data about the flow can these day be acquired even at small amount of the expense relating to experimental (Xia & Sun, 2002). The transient CFD simulations are performed to suggest on the usability of the multiphase approach for the prediction of the solid phase mixing and residence time distribution in the riser (Andreux, et al., 2008). Transient solution is always necessary because of unsteady state nature of fluidized bed. CFD is useful in understanding the quantitative hydrodynamics of fluidization and is needed for the design and scale-up of efficient reactors in several processes industries (Ding & Gidaspow, 1990).

The scientific approach to evaluate the performance of a distributor and overall mixing behavior of a fluidized bed is to measure and analyze the residence time distribution (RTD) of the solid phase (Pant et al., 2014). The gas mixing in circular fluidized bed risers is evaluate as overall behavior. The overall mixing behavior has been studied by measuring the residence time distribution of a gas tracer injected at the feed inlet (Mahmoudi et al., 2010). The knowledge of RTD function is very important requirements for the optimization of the operating parameters and equipment configuration (Idakiev & Morl, 2013). The tracer is introduced at the inlet and monitored at outlet of the column. The concentration of tracer is introduced as a pulse into a fluidized bed column at the feed inlet. Then, the samples are collected at the exit at fixed time intervals until tracer concentration goes to zero at the outlet (Lopez-Isunza, 1975). Using CFD, the flow of inert tracer particles is calculated using species transport equation, where the diffusion coefficient of particles i.e. the necessary parameter indicating particle diffusion capacity is investigated (Hua et al., 2014). Typically, in a CFD study the experimental studies are conducted on a small-scale fluidized bed and numerical codes are used to model the experimental setups and validate the experimental results. The main objective of this work is to uses the CFD code ANSYS FLUENT to simulate the hydrodynamic behavior of a fluidized bed. A first approach of the full-fledged CFD model is the validation of the model with the experimental information. Due to the availability of residence time data, the CFD model is first validated with the RTD information obtained experimentally. The present study is divided into two main approaches: validation of the 2D and 3D simulation of gas-solid fluidized bed with the RTD data and a subsequent study to identify the flow patterns of gas-solid flow in a turbulent fluidized bed. The numerically predicted RTD results are compared with the experimental results of Lopez-Isunza (1975). The target is to make a meaningful comparison between the predictions of the RTD at three different airflow rates that given the normalized velocity $(U_0 / U_{mf} = 9.5; 8.0 \text{ and } 6.4)$. Note that the minimum fluidization velocity is 0.634 cm/s.

2 COMPUTATIONAL FLUID DYNAMICS (CFD) MODEL

2.1 Geometry and mesh

For the CFD simulations, initial mesh needs to be created. This mesh creates a geometry in which the calculations occur and furthermore is divided into several volumes according to the finite-volume methodology. These calculations are governed by the so-called Reynold's averaged Navier-Stokes (RANS) equations. A fluidized bed with a height of 1.5 m and a diameter of 0.2 m is simulated. The particle diameter used in the gas-solid fluidized bed is between 175—200 μm and falls in the Geldart group B particles (Geldart, 1973). The geometry is made in both 2D and 3D. The mesh consists of 12357 elements in 2D and 35616 elements in 3D respectively. This mesh of the fluidized bed is shown by figure 2.1 as follow in 2D and 3D.

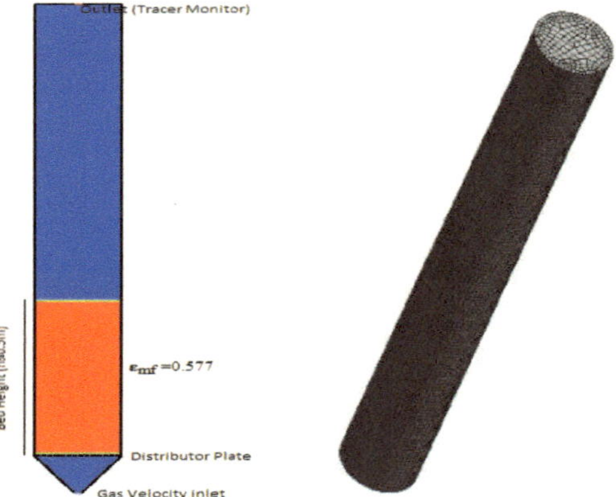

Figure 2.1. Mesh of the Fluidized bed geometry for the study, shown in 2D and 3D respectively.

As seen in figure 2.1, the initial bed height is fixed at 0.5 m. The solid particles can be seen over the distributor plate. The gas inlet is at the bottom and the solids and gas outlet is the top. A tracer monitoring point is also at the top outlet.

2.2. Gas-solid model

For solutions in finite volume, an arrangement of balance equations is numerically solved for fluid flow over a number of control volumes, i.e., over the so-called mesh cells. These balance equations, calculated via the RANS approach for the conservation of mass, momentum and energy were used for CFD simulations. These equations are already implemented in the CFD software ANSYS Fluent® version 16.2. The CFD simulation of the bed hydrodynamics is based on the concept of Eulerian-Eulerian two phase flow model incorporated in the kinetic theory of granular flow (KTGF) (Gidaspow et al, 2004; Tartan & Gidaspow, 2004). The kinetic theory of granular flow (KTGF) originating from the theory for non-ideal dense gases developed by (Chapman & Cowling, 1970) is used to close the solid phase equations. A brief explanation of the underlying theory related to KTGF can be found in Section 2.2.3.

2.2.1 Conservations of mass

The sum of volume fraction for gas and solid phases must equal to unity:

$$\alpha_g + \alpha_s = 1 \tag{1}$$

where a subscript *g* and *s* stand for gas and solid phases.

The continuity equation for a gas phase and a solid phase are express as following:

$$\begin{gathered} \frac{\partial}{\partial t}(\alpha_g \rho_g + \nabla.(\alpha_g \rho_g \vec{v}) = 0 \\ \frac{\partial}{\partial t}(\alpha_s \rho_s + \nabla.(\alpha_s \rho_s \vec{v}) = 0 \end{gathered} \tag{2}$$

2.2.2 Conservation of Momentum

The momentum equation for the gas (g) phase as given as follow:

$$\frac{\partial}{\partial t}(\alpha_g \rho_g \vec{v_g}) + \nabla.(\alpha_g \rho_g \vec{v_g} \vec{v_g}) = -\alpha_g \nabla p + \nabla \overline{\overline{\tau_g}} + \alpha_g \rho_g \vec{g} - K_{gs}(\vec{v_g} - \vec{v_s}) \tag{3}$$

Momentum equation for solid phase:

$$\frac{\partial}{\partial t}(\alpha_s \rho_s \vec{v_s}) + \nabla.(\alpha_s \rho_s \vec{v_s} \vec{v_s}) = -\alpha_s \nabla p - \nabla p_s + \nabla \overline{\overline{\tau_s}} + \alpha_s \rho_s \vec{g} - K_{gs}(\vec{v_g} - \vec{v_s}) \tag{4}$$

Gas phase shear stress tensor

$$\overline{\overline{\tau_g}} = \alpha_g \mu_g (\nabla \vec{v_g} + \overrightarrow{v_g^T}) \tag{5}$$

Solid phase shear stress tensor

$$\overline{\overline{\tau_s}} = \alpha_s \mu_s(\nabla \vec{v}_s + \vec{v}_s^T) + \alpha_s(\lambda_s + \frac{2}{3}\mu_s)\nabla \vec{v}_s. \quad 6$$

The solids phase stress tensor consists a shear viscosity and a bulk viscosity, and arising from solid particle momentum exchange due to kinetics and collision. A frictional viscosity is taking into account for viscous-plastic translation, it is occurring when a solid phase is reaching the maximum volume fraction. The sum of collisional, kinetic and frictional viscosities to give the solids shear viscosity ($\mu_s = \mu_{s,\,col} + \mu_{s,\,kin} + \mu_{s,\,fr}$).

Solid's collision viscosity: Syamlal et al. (1993)

$$\mu_{s,\,col} = \frac{4}{5}\alpha_s \rho_s d_p g_{o,\,ss}(1 + e_{ss})(\Theta)_s^{1/2} \quad 7$$

Solids kinetic viscosity: Syamlal et al. (1993)

$$\mu_{s,\,kin} = \frac{\alpha_s \rho_s d_p}{6(3 - e_{ss})}\sqrt{\pi\Theta_s}\left[1 + \frac{2}{5}\alpha_s g_{o,\,ss}(1 + e_{ss})(3e_{ss} - 1)\right] \quad 8$$

Solids frictional viscosity: Schaeffer (1987)

$$\mu_{s,\,fr} = \frac{p_s}{2\sqrt{I_{2D}}}\sin\varphi \quad 9$$

where φ is the angle of internal friction.

Solids bulk viscosity: Lun et al. (1984)

$$\lambda_s = \frac{4}{3}\alpha_s \rho_s d_p g_{o,\,ss}(1 + e_{ss})(\frac{\Theta}{\pi})_s^{1/2} \quad 10$$

Solid pressure is implemented according to Lun-et-al (1984)

$$p_s = \alpha_s \rho_s \Theta_s + 2\rho_s g_{o,\,ss}(1 + e_{ss})\alpha_s^2 \Theta_s \quad 11$$

Radial distribution function is implemented according to Ogawa et al. (1980)

$$g_{o,\,ss} = \left[1 - (\frac{\alpha}{\alpha_{s,\,\max}})^{1/3}\right]_s^{-1} \quad 12$$

where the restitution coefficient for particle-particle collisions, $\boldsymbol{e}_{ss}$ is fixed at 0.9 for all the simulations in the present study. This value is taken from literature studies (Balakin et al. 2012; Hu et al., 2010) for glass and polymer particles (in this study polymeric silica gel is used).

Wen-Yu and Gidaspow drag models are used for the momentum exchange coefficient. The Gidaspow drag model equation is a combination of Wen-Yu and Ergun equations.

Drag function: Wen-Yu model (1966)

$$K_{gs} = \frac{3}{4} C_D \frac{\alpha_g \alpha_s \rho_g}{d_p \alpha_g^{2.65}} \left| \vec{v}_g - \vec{v}_s \right| \qquad 13$$

Drag function: Gidaspow model (1994)

$$K_{gs} = 150 C_D \frac{\alpha_s \mu_g}{\alpha_g d_p^{2.}} (1 - \alpha_g) + 1.75 \frac{\alpha_g \rho_g}{d_p} \left| \vec{v}_g - \vec{v}_s \right| \text{ for } \alpha_g \leq 0.8 \qquad 14$$

when $\alpha_g \geq 0.8$, Gidaspow drag model becomes Wen-Yu model.

$$C_D = \frac{24}{\alpha_g \mathrm{Re}_s} \left[1 + (0.15 \alpha_g \mathrm{Re}_s)^{0.687} \right] \text{ and } \mathrm{Re}_s = \rho_g \frac{d_p}{\mu_g} \left| \vec{v}_g - \vec{v}_s \right|.$$

2.2.3 Kinetic theory granular flow (KTGF)

The granular temperature is computed by solving a fluctuating kinetic energy equation for the particles, following the Kinetic Theory of Granular Flow (KTGF) principle. To specify collisional energy dissipation, γ_s, due to inelastic collisions of particles and the granular conductivity, K_s, the equation is given as:

$$\frac{3}{2} \left[\frac{\partial}{\partial t} (\alpha_s \rho_s \Theta) + \nabla \alpha_s \rho_s \Theta \right] = \bar{\tau} : \nabla \vec{V}_s + \nabla (K_s \nabla \Theta - \gamma_s) \qquad 15$$

It is $\Theta = \frac{1}{3} V_s^2$ is the granular temperature.

A list of sub-models used for the CFD model and their values are found in Table 1. For simulations of these equations inside the fluidized bed, various boundary conditions are required. The simulations in this study are with gas phase (air) with density 1.225kg/m^3 and viscosity 1.785x10^{-5} Pa s and solid phase of particle diameter between 175-200μm, with density 2400 kg/m^3. For tracer gas (Helium) density and viscosity 0.165 kg/m^3 and 2.0x10^{-5} Pa s respectively.

Table 1. Summary for the numerically simulation boundary conditions and model parameters

Descriptions	Sub-model
Granular viscosity	Syamlal O'Brien (2003)
Granular bulk viscosity	Lun et al. (2009)
Frictional viscosity	Schaeffer (1987)
Granular temperature	KTGF (Gidaspow, 2004)

Drag law	Wen-Yu (1966) and Gidaspow (1994)
Inlet boundary condition	Velocity inlet
Outlet boundary condition	Pressure outlet
Wall boundary condition	No slip for air
Parameters	Values
Bed height	0.5 m
Volume fraction of solid phase	0.577
Operating pressure	101325 Pa
Gas inlet velocity	(0.065, 0.051 and 0.041) m/s
Turbulent kinetic energy	0.30554 m^2/s^2
Turbulent dissipation rate	0.30553 m^2/s^3
Angle of internal friction	30°
Particle-particle restitution coefficient	0.9
Specularity coefficient for solid phase	0.001

2.3 Gas-solid mixing studies

Gas-solid fluidization is divided into several regimes depending on their gas velocity. In this study, the gas velocity used falls within bubble fluidized bed regime, that is the range of Reynold number (Re) between 0.2 and 1000. Gas bubbles rising in a fluidized bed continuously transfer gas to the dense phase by diffusion and convection mechanisms. The gas that enters the fluidized bed is passed through the granular particles in the column. The bubble pushes through the bed of solid particles and leave at the top. Thus, around the bubble, there is the cloud of dense phase, which the gas is continuously in convective exchange with gas inside the bubble. Inside the cloud, interaction of solid and gas takes place. This implies that gas is present in the cloud which in principle flows back to the bubble again and part of it may be transferred to the dense phase by diffusion or adsorption with the solids that is exchanged with the emulsion phase. Convection and diffusion mechanisms of exchange occur simultaneously. The gas interchange between bubble and emulsion phases has been described in terms of mass transfer coefficient by Kunii & Levenspiel (1991):

$$\frac{1}{\text{intrfacial area}}\frac{dN_A}{dt} = k\left(C_{Ab} - C_{Ae}\right) \quad 16$$

where k is mass transfer coefficient (cm/s). As interchange rate is defined as

$$\frac{1}{volume}\frac{dN_A}{dt} = k\left(C_{Ab} - C_{Ae}\right) \quad 17$$

where k is mass transfer coefficient (s^{-1}).

Kunii and Levenspiel (1991) assumed that there are two transfer steps, namely the transfer between bubble void and cloud particles overlap region and that between the cloud particles overlap region and the emulsion phase. They further assumed that the mass transfer coefficient for bubble-cloud and cloud-emulsion given as in terms of gas exchange coefficient per unit volume of bubble void.

$$k_{bc} = 4.5\frac{u_{mf}}{D_b} + 5.85\left(\frac{D^{1/2}g^{1/4}}{D_b^{5/4}}\right) \quad \text{for bubble to cloud} \quad 18$$

$$k_{ce} = 6.78\left(\frac{\alpha_{mf}u_b D_e}{D_b^3}\right)^{1/2} \quad \text{for cloud to emulsion} \quad 19$$

They defined the overall gas exchange coefficient between bubble phase and emulsion phase as following:

$$k_b = \frac{k_{bc}k_{ce}}{k_{bc} + k_{ce}} \quad 20$$

From the detailed information concerning the mechanics of fluid (gas) in the cloud, it can safely be assumed that the gas composition in the cloud is approximately uniform. Chiba and Kobayashi (1970) used the analysis by Murray for the flow pattern around the spherical bubble to obtain the following expressions for the interchange coefficient:

$$k_b = \frac{6.78}{1-\alpha_g}\left(\frac{D\alpha_{mf}u_b}{D_b^3}\right) \quad 21$$

Finally, Kobayashi et al. (1967) determined the rate of exchange by measuring the RTD in a fluidized bed. They concluded that the interchange parameter could be expressed as:

$$k_b = \frac{11}{D_b}. \quad 22$$

Note that eq. (23) indicates no effect of the diffusion coefficient of the transferring species, nor any direct effect of the fluid mechanical phenomenal occurring. However, in this study, Gidaspow and Wen-Yu gas-solid interchange coefficients are used to observe the behavior of the hydrodynamics in a fluidized bed.

2.3.1 Species transport model

Species transport equation is used for the residence time distribution (RTD) study. The conservation equation for the species predicts the local mass fraction of each species, Y_i, through the solution of a convection and diffusion equation from i^{th} species. The general conservation equation is given as following:

$$\frac{\partial}{\partial t}(\rho_i)+\nabla.\left(\rho\vec{v}Y_i\right)=-\nabla\vec{J}_i+R_i+S_i \qquad 23$$

where ρ_i is density of species i, $\vec{v}$ is velocity vector, $\vec{J}i$ is diffusion flux of species i, R_i is the net rate of production of species by chemical reaction and S_i is the rate of creation by addition from the dispersed phase. In this case, chemical reaction is not considered, therefore R_i and S_i are ignored and (Eq.24) is reduced to:

$$\frac{\partial}{\partial t}(\rho_i)+\nabla.\left(\rho\vec{v}Y_i\right)+\nabla\vec{J}_i=0 \qquad 24$$

where the diffusion flux of the species for turbulent flow is calculated by Fick`s law and Ansys Fluent® uses the dilute approximation of Fick`s law to model mass diffusion due to concentration gradients and temperature and the diffusion flux can be written as:

$$\vec{J}_i=\nabla.Y_i\left(\rho D_{i,m}+\frac{\mu_t}{S_{ct}}\right)-D_{ti}\frac{\nabla T}{T} \qquad 25$$

where S_{ct} is the turbulent Schmidt number $S_{ct}=\frac{\mu_i}{\rho D_t}$ while the temperature gradient is negligible and combine the Equation (25 and 26) form given as:

$$\frac{\partial}{\partial t}(\rho_i)+\nabla.\left(\rho\vec{v}Y_i\right)=\nabla.Y_i\left(\rho D_{i,m}+\frac{\mu_t}{S_{ct}}\right). \qquad 26$$

During the simulation, diffusion at the inlet is important because the pressure-based solver in calculates the net transport of species at inlets consists of both convection and diffusion components. That means the convection component fixed by the specified inlet species mass fraction, while the diffusion component is depends on the gradient of the computed species concentration field (which is not known priori).

2.3.2 Turbulent Modeling

In various industrial processes related to fluidization, turbulence behavior is observed. k -ε model is the most widely used engineering turbulent model for industrial applications, robust and reasonable accurate.

In this study, a standard k- ε model is use to solve the transport equation for turbulent kinetic energy (k) and dissipation rate of the turbulent kinetic energy (ε). The k- ε model written as follows:

$$\nabla.(\rho_m k\vec{v}_m) = \nabla.(\frac{\mu_{t,m}}{\sigma_\varepsilon}\nabla k) + G_{k,m} - \rho_m \varepsilon \qquad 27$$

$$\nabla.(\rho_m k\vec{v}_m) = \nabla.(\frac{\mu_{t,m}}{\sigma_\varepsilon}\nabla k) + \frac{\varepsilon}{k}(c_{1\varepsilon}G_{k,m} - c_{2\varepsilon}\rho_m \varepsilon) \qquad 28$$

where $\mu_{t,m}$ is turbulent (eddy) viscosity it's computed from k – ε, ($\mu_{t,m} = \rho_m C_\mu \frac{k^2}{\varepsilon}$)

where C_μ=0.09, C_{e1}=1.44, C_{e2}=1.92, σ_ε=1 and $G_{k,m}$ is generation of turbulent kinetic energy due to velocity gradient

2.4 Residence time distribution (RTD)

Residence time distribution (RTD) study is very important to characterize mixing and flow patterns inside a reactor, and to know whether the unit is approaching an ideal reactor i.e., plug flow reactor or mixed flow reactor. Besides, RTD is used to model the reactor as a combination of ideal reactors. Residence time distribution data values can be used to investigate any non-idealities like channeling, by passing and short circuiting present in the a reactor (Levenspiel, 1999). It can also be used to scale up or to design a reactor once the kinetics is obtained. To accomplish RTD simulations, the transient analysis of a tracer is considered with the same physical properties as that of continuous phase. The transport equation for the concentration of a turbulent flow can written as follow:

$$\frac{\partial}{\partial t}(\rho_m Y_{tr}) + \frac{\partial}{\partial x_j}(\rho_m \vec{v}_m Y_{tr}) = \frac{\partial}{\partial x_j}[\frac{\partial}{\partial x_j} Y_{tr}\left(\rho D_{i,m} + \frac{\mu_t}{S_{ct}}\right)] \qquad 29$$

Where μ_t is a turbulent viscosity and given as $\mu_t = \rho_m C_\mu \frac{k^2}{\varepsilon}$, and Y_{tr} is mass fraction of tracer species.

2.5 Numerical Procedure for hydrodynamics

2.5.1 Numerical Approach

ANSYS Fluent® solves the governing integral equations for conservation mass, momentum, kinetic theory of granular flow and turbulent simultaneously. The technique uses a finite volume approach for flow arrangements.

2.5.2 Pressure-Based Solver Algorithm

A pressure-based solver employs phasic momentum equations, shared pressure, and phasic volume fractions equations in a segregated way. The phase coupled implicitly technique for pressure connected equations (PC-SIMPLE) algorithm is used, which is an extension of the SIMPLE algorithm, developed for multiphase flows. Pressure and velocity corrections are then applied to satisfy the volume continuity constraint.

2.5.3 Spatial Discretization

A finite volume method, capable of handling complex geometries is applied. The discretization of the convective fluxes is first order upwind. Under transient formulation, the bounded second order implicit is used; this formulation would provide better stability, since time discretization would always ensure the bound for variables if available. The bounded second order implicit is better option for Fluidized bed to obtain stability and accurate results. A first order implicit unsteady formulation using the phase-Couple SIMPLE algorithm (Syamlal et al., 1998) is used to solve the equations for both two-dimensional and three-dimensional cases.

2.5.4 Initial and Boundary Conditions

Initial condition may not influence the steady state solutions that is desired in fluidized bed modeling. Notwithstanding, strategically chosen initial conditions help to ensure the convergence of the solution. There are two types of initial conditions: solids volume fraction in the bed and Y-velocity of the gas phase in fluidized bed. The solids volume fractions in the freeboard is initially set to zero (assuming only gas). They Y-velocity of the gas phase in the fluidized bed is computed through a steady state volumetric flow rate balance in which the flow rate entering a fluidized bed segment is linked to the flow rate leaving the fluidized bed segment (having both solids and gas phase):

$$v_{in}u_{in} = v_g u_g \qquad 30$$

$$u_g = \frac{v_{in} u_{in}}{v_g} \quad 31$$

Note that the situations represent the superficial velocity of the gas where the proportion of the bed volume to the inlet volume is identical to the void fraction as shown in:

$$u_g = \frac{u_{in}}{\alpha_g} \quad 32$$

Boundary conditions: at inlet of the fluidized bed is gas superficial velocity for primary phase, while the secondary phase is zero inlet velocity. At outlet of the fluidized bed is atmospheric pressure. No slip, wall boundary condition is applied to the gas phase, while the partial slip conditions is applied to the solid phase. A specualarity coefficient ($\phi = 0.001$) for a partial slip model is used following the suggestion of Shi et al. (2015).

3 Problem Description and CFD Method

3.1 Problem description

A schematic conceptual representation of the Fluidized bed is shown in fig. 2.1. The figure represents a cylindrical vessel i.e., the gas-solid fluidization bed with superficial gas velocity at the bottom inlet of the bed. The solids particles are patched with α_s =0.577 at initial bed height (h_o) = 0.5 m. The fluidizing gas causes the gas-solids mixing. The gravitational effect on the particles bed is taken into accounted as the bed is vertical.

3.2 Simulation procedure

The assumption that a tracer does not undergo any chemical reaction with the reactants or products is valid. The tracer amount is very small and all the physical properties of the tracer are practically same as the working fluid. This way the flow is not disturbed inside the reactor after the tracer is introduced. This assumption is used in many literature studies (Hua et al., 2014; Shilapuram, et al., 2011). The RTD characteristics such as mean residence time, variance, and exit age distribution are possible to calculate.

The tracer gas injection method used:

1. Initially, run the steady state simulations for the flow rate and turbulence as provided in table 1, for physical properties and boundary conditions. It may be noted that the residuals of the continuity, momentum, k-turbulent kinetic, ε - dissipation rate kinetic energy reduce on reaching steady state conditions.
2. Once the convergence is achieved for the steady state simulation, the mass fraction of the tracer is made to 1 and the other equations disabled. The tracer carries the properties of Helium gas, as mentioned in the experimental setup (see Lopez-Isunza, 1975).
3. Now the transient equation for the tracer (Eq. (29)) is carried out for one iteration to imitate the Pulse method after which the mass fraction of the tracer should be changed to zero.
4. The flow equations (continuity, momentum, k-turbulent kinetic, ε - dissipation rate kinetic energy species transport and tracer gas) are solved.
5. Then concentration of the tracer is time monitored at the exit of the fluidized bed.

4 RESULTS AND DISCUSSION

CFD simulations of the fluidized bed are performed and a RTD validation study is done by comparing the results with the experimental study of Lopez-Isunza (1975).

4.1 Gas-Solid hydrodynamics

Hydrodynamic study of gas-solid fluidized bed simulated using CFD in 2D with the inclusion of drag models of Gidaspow and Wen-Yu has been performed. Figure 4.1 represents the consideration of time-averaged solids velocity profile in the radial direction at height of 0.3m. The motoring height is chosen arbitrarily but care is taken to ensure that it is below the top height of the fluidized bed. This is done to ensure that the gas-solids mixing behavior is certainly investigated. It can be seen that the simulation results using Gidaspow and Wen-Yu gas-solid drag models are slightly different from each other. The solids axial velocity fluctuations along the radial position are larger in Wen-Yu model than in Gidaspow model. Experimental studies (Li et al., 2009; Loha et al., 2012) indicate that the variations in solids velocity is similar to that exhibited by Gidaspow drag model. This indicates that the Gidaspow model could be more accurate in predicting the gas-solid solid hydrodynamics in this study. Moreover, Gidaspow drag equations is a modified form of Wen-Yu (and Ergun) drag equations as explained above (section 2.2.2).

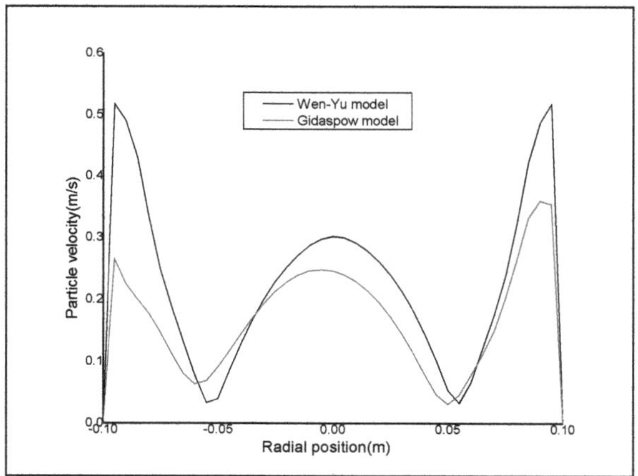

Figure 4.1a. Time-averaged (10 s) simulation of solid particles axial velocity along the radial height of 0.3m.

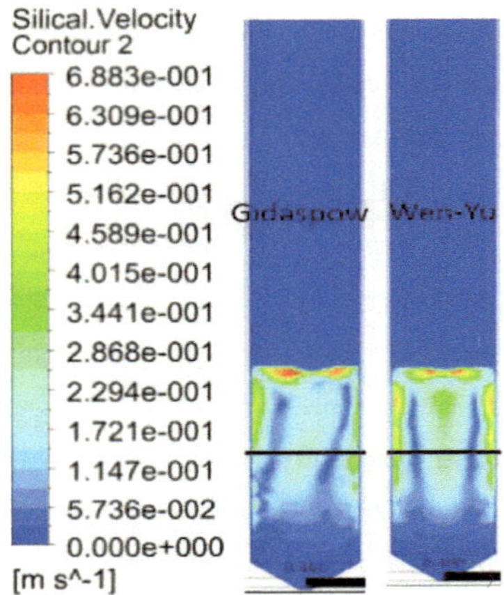

Figure 4.1 b. Time-averaged (10 s) simulation of solid particles velocity contour

Fig. 4.2 illustrates time-averaged solid volume fraction variation along the center of the fluidized bed. Fig. 4.2 shows a maximum solids volume fraction in the model of Wen-Yu ($\alpha_s = 0.63$) compared to the model of Gidaspow ($\alpha_s = 0.59$).

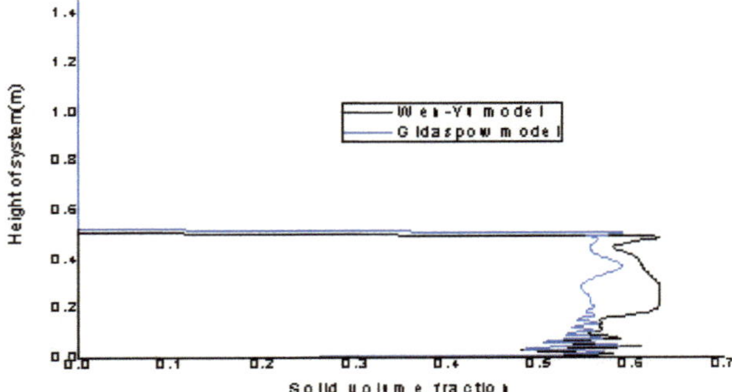

Figure 4.2 a. Time-averaged (10 s) simulation of solid particle volume fraction changed along the center axis of the fluidized bed.

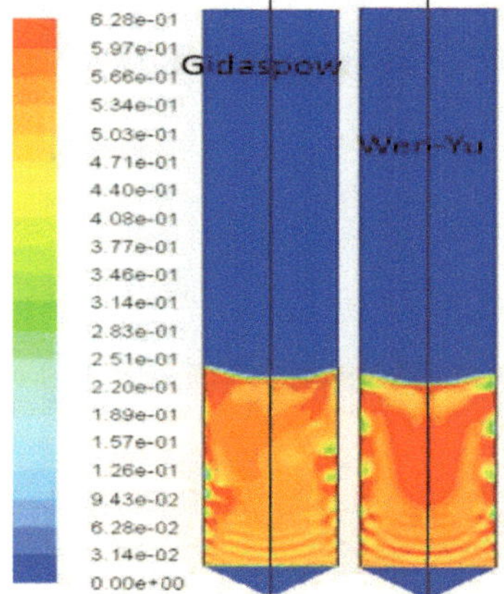

Figure 4.2 b. Time-averaged (10 s) simulation of solid particle volume fraction contour in the fluidized bed.

This indicates that the Wen-Yu model promotes solids clustering as compared to the Gidaspow model. This is probably due to the high gas velocity escaping along the walls thereby pushing the solid particle together and forming cluster. Figure 4.3 shows the profile of time-averaged solid volume fraction profile in the radial position at height of 0.4m by Gidaspow and Wen-Yu drag models.

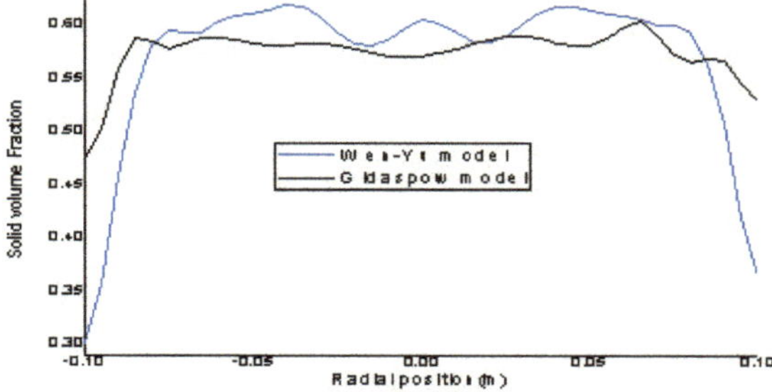

Figure 4.3 a. Time-averaged (10 s) simulation of solid volume fraction change along the radial position at a height of 0.4m.

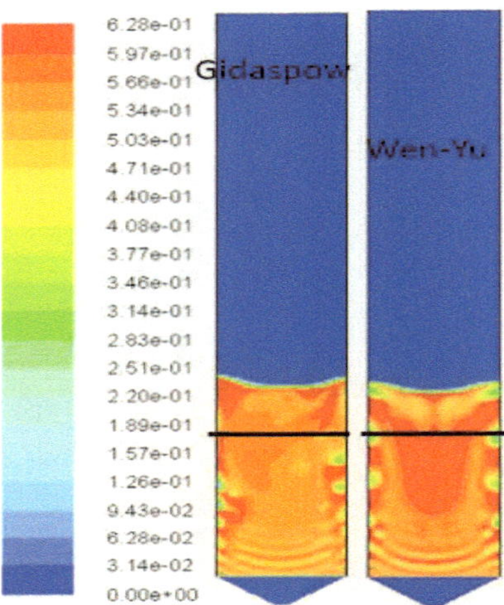

Figure 4.3 b. Time-averaged (10 s) simulation of solid particle volume fraction contour in the fluidized bed.

The solids volume fraction shows even distribution along the radial direction for both the models, while Wen-Yu model indicates comparatively larger variation of solids volume fraction along the radial direction as compared to Gidaspow model. To conclude from the above three figures, Gidaspow model seems to be preferable than We-Yu model as it indicates less variation along the radial and axial direction in gas and solids velocity. Also literature studies (Sahoo & Sahoo et al., 2015; Loha et al. 2012; Hooyar et al., 2012) appreciate Gidaspow model for a comparison with experimental data. It must also be noted that the Gidaspow model is modification of Ergun and Wen-Yu models.

4.2 Residence Time Distribution (RTD) validation

The species transport model in ANSYS Fluent® used for the simulation of residence time distribution (RTD) is implemented for the fluidized bed in 2D and 3D respectively. In the RTD study, Helium gas considered as tracer gas. The concentration of the gas is measured at the exit and the residence time characteristics calculated as shown below.

Mean time distribution: $\bar{t} = \frac{\sum c_i t_i \Delta t_i}{\sum c_i \Delta t_i}$ 33

Exit time distribution: $E(t) = \frac{c(t)}{\sum c_i \Delta t_i}$ 34

Three cases with air velocities ranging from 6.4-9.5 times the minimum fluidization velocity is simulated in this study. Note that the minimum fluidization velocity U_0 / U_{mf} is 0.634 cm/s. In all the three cases $(U_0 / U_{mf} = 9.5; 8.0 \text{ and } 6.4)$, both the 2D and 3D simulation results agree quite well with the experimental data (figs. 4.4-4.6 respectively) especially at lower simulation times while it seem to over-predict the experimental data as the simulation time proceeds in the end. This is probably because of the numerical discrepancies associated with the CFD model while obtaining the final tracer concentration at the outlet. However, the peaks from the experimental data fit well with the numerical model, although the peaks from the 2D simulations are slightly lower than the peaks from the 3Dsimualtions. This is probably because in 3D, tracer has enough space to passing through the fluidized bed.

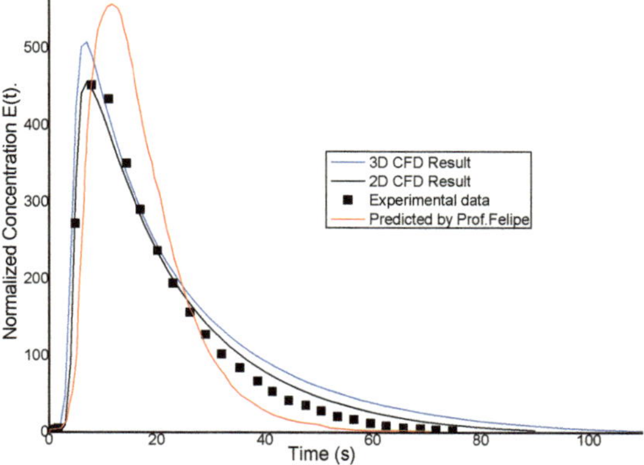

Figure 4.4. Experimental and predicted RTD response of the fluidized bed under normalized velocity conditions $(U_0 / U_{mf} = 9.5)$

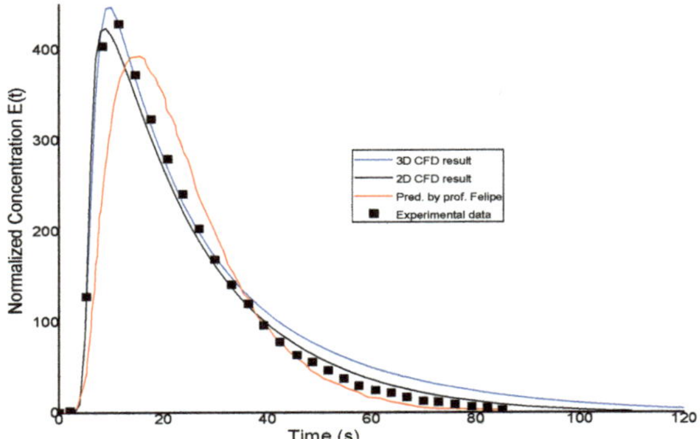

Figure 4.5. Experimental and predicted RTD response of the fluidized bed under normalized velocity conditions ($U_0 / U_{mf} = 8.0$)

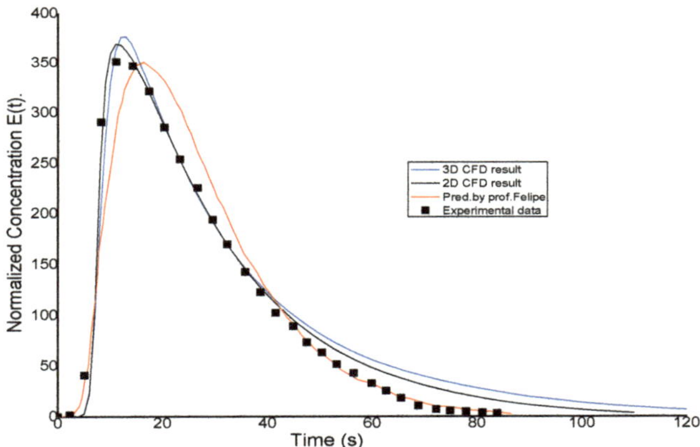

Figure 4.6. Experimental and predicted RTD response of the fluidized bed under normalized velocity conditions ($U_0 / U_{mf} = 6.4$)

The validation of residence time distribution studies seems to be in agreement with the experimental study of Lopez-Isunza (1975) for the three different velocity conditions investigated (refer to Figs. 4.4-4.6). The theoretical mean residence time of the gas in beds is given by Kunii & Levenspiel (1991) as $\bar{t} = \frac{\alpha_g h_g}{U_0}$. The results are summarized in Table 2 for an initial solids bed height of 0.5 m.

Fluidized bed column height, H (m)	Normalized velocity	Mean residence time (s) (Lopez-Isunza, 1975)		Theoretical mean residence time (s) (Levenspiel, 1991)	CFD mean residence time (s)	
		Experiment	Predicted		2D	3D
1.5	9.5	19	14.05	20	22	24.13
1.5	8.0	23.297	18.796	25.490	25.3	28.8
1.5	6.4	25.402	22.320	31.707	29.5	33

Table 2. Summary of the numerical results of the mean residence time, compared with the results from Lopez-Isunza (1975) and Levenspiel (1991).

For $U_0 / U_{mf} = 9.5$, the mean residence time is the fastest (compared to $U_0 / U_{mf} = 6.4$ which is the slowest) which means that the mass transfer between the bubbles and emulsions phase (see Section 2.3) is comparatively faster.

5 CONCLUSION

The gas-solid hydrodynamics using computational fluid dynamics (CFD) is important to identify the flow behavior inside the fluidized bed. The CFD model implemented in this work is an Eulerian-Eulerian multiphase flow coupled with the kinetic theory of granular flow and standard turbulence model used for closure. The two drag models, necessary for gas-solid interchange coefficient i.e., the Gidaspow and Wen-Yu models, simulate the gas-solids mixing in the bed. Although there is not much difference between the performance of these two models, Gidaspow model is recommended as it is able to correctly capture the hydrodynamics inside the fluidized bed. The simulations results studied are time-averaged distributions of gas and solids velocity along the axial and radial directions. The simulation results related to residence time distribution (RTD) for three different velocity rations ($U_0 / U_{mf} = 9.5; 8.0$ and 6.4) i.e., at different degrees of gas-solids mixing is evaluated in both 2D and 3D. The results are in general agreement with the experimental data of Lopez-Isunza (1975). However, further research is still needed to understand the mass transport between the bubble and emulsion phases and connect them to the gas-solid hydrodynamics. It is important to investigate the relation between the mass transport and the residence time distribution, in order to have a better understanding of the fluidized bed reactor.

6 References

Kunii, Daizo & Octave Levenspiel. (1991). *Fluidization Engineering*. 2. ed. Butterworth-Heinemann Series in Chemical Engineering. Boston: Butterworth-Heinemann.

Cui, H., & Grace, J. R. (2007). Fluidization of biomass particles: A review of experimental multiphase flow aspects. *Chemical Engineering Science*, *62*(1–2), 45–55.

Gidaspow, D. (1994). *Multiphase flow and fluidization: continuum and kinetic theory descriptions*. Boston: Academic Press.

Salaices, E., de Lasa, H., & Serrano, B. (2012). Steam gasification of a cellulose surrogate over a fluidizable Ni/α-alumina catalyst: A kinetic model. *AIChE Journal*, *58*(5), 1588–1599.

Tartan, M., & Gidaspow, D. (2004). Measurement of granular temperature and stresses in risers. *AIChE Journal*, *50*(8), 1760–1775.

Richardson, J. F., Harker, J. H., Coulson, J. M., & Richardson, J. F. (2008). *Particle technology and separation processes* (5. ed.). Oxford: Butterworth Heinemann.

Pant, H. J., Sharma, V. K., Goswami, S., Samantray, J. S., Mohan, I. N., & Naidu, T. (2014). RTD in a pilot-scale gas–solid fluidized bed reactor using radiotracer technique. *Journal of Radioanalytical and Nuclear Chemistry*, *302*(3), 1283–1288.

Xia, B., & Sun, D.-W. (2002). Applications of computational fluid dynamics (CFD) in the food industry: a review. *Computers and Electronics in Agriculture*, *34*(1), 5–24.

Idakiev, V., & Morl, L. (2013). Study of Residence Time of Disperse Materials in Continuously Operating Fluidized Bed Apparatus. *Journal of Chemical Technology and Metallurgy*, *48*(5), 451–456.

Lopez-Isunza, F. (1975). *Determination of interchange parameter in fluidized beds using pulse testing techniques*. M.Eng Thesis, McMaster University, Canada.

Benzarti, S., Mhiri, H., & Bournot, H. (2012). Drag models for simulation gas-solid flow in the bubbling fluidized bed of FCC particles. *World Academy of Science, Engineering and Technology*, *61*, 1138–1143.

Taghipour, F., Ellis, N., & Wong, C. (2005). Experimental and computational study of gas–solid fluidized bed hydrodynamics. *Chemical Engineering Science*, *60*(24), 6857–6867.

Wen C.-Y and Yu Y. H. *Mechanics of Fluidization*. Chem. Eng. Prog. Symp. Series. 62. 100–111. 1966.

Dutta, A., Ekatpure, R. P., Heynderickx, G. J., de Broqueville, A., & Marin, G. B. (2010). Rotating fluidized bed with a static geometry: Guidelines for design and operating conditions. *Chemical Engineering Science*, *65*(5), 1678–1693.

Andreux, R., Petit, G., Hemati, M., & Simonin, O. (2008). Hydrodynamic and solid residence time distribution in a circulating fluidized bed: Experimental and 3D computational study. *Chemical Engineering and Processing: Process Intensification*, *47*(3), 463–473.

Khongprom, P., Aimdilokwong, A., Limtrakul, S., Vatanatham, T., & Ramachandran, P. A. (2012). Axial gas and solids mixing in a down flow circulating fluidized bed reactor based on CFD simulation. *Chemical Engineering Science*, *73*, 8–19.

Li, P., Lan, X., Xu, C., Wang, G., Lu, C., & Gao, J. (2009). Drag models for simulating gas–solid flow in the turbulent fluidization of FCC particles. *Particuology*, *7*(4), 269–277.

Hua, L., Wang, J., & Li, J. (2014). CFD simulation of solids residence time distribution in a CFB riser. *Chemical Engineering Science*, *117*, 264–282.

Shi, X., Sun, R., LAN, X., Liu, F., Zhang, Y., & GAO, J. (2015). CPFD simulation of solids residence time and back mixing in CFB risers. *Powder Technology*, *271*, 16–25.

Chapman, S., & Cowling, T. G. (1970). *The mathematical theory of non-uniform gases: an account of the kinetic theory of viscosity, thermal conduction and diffusion in gases*. Cambridge University Press.

Loha, C., Chattopadhyay, H., & Chatterjee, P. K. (2012). Assessment of drag models in simulating bubbling fluidized bed hydrodynamics. *Chemical Engineering Science*, *75*, 400–407.

Kobayashi, H., Arai F. & Sunagawa T. (1967). Kagaku Kogaku (Chemical Engineering., Japan) 29, 858.

Chiba T. & Kobayashi H. (1970). Gas exchange between the bubble and emulsion phases in gas-solid fluidized beds. *Chemical Engineering science*, *25*(9), 1375-1385.

Balakin, B., Hoffmann, A. C., & Kosinski, P. (2012). The collision efficiency in a shear flow. *Chemical Engineering Science*, *68*(1), 305–312.

Hu, X., Zhang, J., Dong, C., & Yang, Y. (2010). Simulation of Dense Gas-Solid Fluidized Bed Based on Two-Fluid Model. In *2010 Asia-Pacific Power and Energy Engineering Conference* (pp. 1–4). IEEE.

Syamlal, M., Rogers, W. & O'Brien, T.J. (1993). *MFIX documentation.* National Energy Technology Laboratory, Department of Energy, Technical Note No. DOE/MC31346-5824.

Geldart, D. (1973). Types of gas fluidization. *Powder Technology*, *7*(5), 285–292.

Levenspiel, O. (1999). *Chemical reaction engineering* (3rd ed). New York: Wiley.

Hua, L., Wang, J., & Li, J. (2014). CFD simulation of solids residence time distribution in a CFB riser. *Chemical Engineering Science*, *117*, 264–282.

Cizmas, P. G., Palacios, A., O'Brien, T., & Syamlal, M. (2003). Proper-orthogonal decomposition of spatio-temporal patterns in fluidized beds. *Chemical Engineering Science*, *58*(19), 4417–4427.

Gidaspow, D., Jung, J., & Singh, R. K. (2004). Hydrodynamics of fluidization using kinetic theory: an emerging paradigm. *Powder Technology*, *148*(2–3), 123–141.

Schaeffer, D. G. (1987). Instability in the evolution equations describing incompressible granular flow. *Journal of Differential Equations*, *66*(1), 19–50.

Shuyan, W., Xiang, L., Huilin, L., Long, Y., Dan, S., Yurong, H., & Yonglong, D. (2009). Numerical simulations of flow behavior of gas and particles in spouted beds using frictional-kinetic stresses model. *Powder Technology*, *196*(2), 184–193.

Sahoo, P., & Sahoo, A. (2015). A Comparative Study on Effect of Different Parameters of CFD Modeling for Gas–Solid Fluidized Bed. *Particulate Science and Technology*, *33*(3), 273–289.

Mottaghi, M., Attarb, H., & Torabzadegana, M. (2012). *Assess proper drag coefficient models predicting fluidized beds of CLC reactor by utilizing computational fluid dynamic simulation.* 53rd Scandinavian conference on simulation and modeling, organized, Reykjavík, Iceland.